Ogi Orange
THE BIG RACE

ISBN 979-8-9875895-2-6

Published by Pybabi Publishing

This book is dedicated to my loving in-laws. I thank you for your support and for the privilege to be your son.

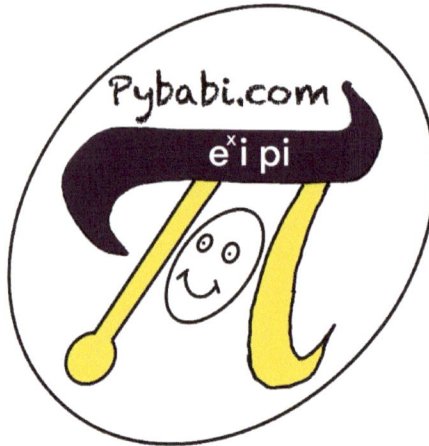

Message for instructor, teacher, or parent:

This book is best used by allowing one of the children who can read to lead the group after you have done it first as a model for them to imitate. Then they should read the story or section, ask the questions in the MATCH and ANSWER sections to their fellow students or if in a family to their brothers and sisters. As they lead the group you can monitor their progress. As the monitor you should answer any questions that arise naturally from the discussion. After one child has taken the lead, the role of teacher should be rotated to another child and then repeated until each child has successfully taught the material before advancing to another book. Since they are learning to teach the material they will naturally pay sufficient attention to learn the material and the lessons will stick with them. As the monitor, it may be best to intervene only as necessary allowing the children to solve and figure things out on their own until they need assistance.

If the child cannot read the parent can read and explain the lesson being taught. As soon as the child can understand even the simplest principle or idea have them to explain it back to you as if you are the student and they are the teacher.

When children feel they have an important role in education they will naturally learn what is necessary to play that role.

Of course this is not the only way to use the material, as you help the student you will soon discover which methods are best depending on the age and ability of the child. Please note that the illustrations in the book are specifically designed not to be perfect. They are only the beginning of a bigger idea or concept that will later be explained in greater detail. By keeping this in mind you will not have to push the student into trying to understand everything at the same time.

Thank you for choosing to use this material.

pybabi.com

OGI ORANGE

THE BIG RACE

Part I

Written and Illustrated by Lee Chau

OGI ORANGE

THE BIG RACE

Part II

Written and Illustrated by Lee Chau

OGI ORANGE

THE BIG RACE

Part III

Written and Illustrated by Lee Chau

OGI ORANGE

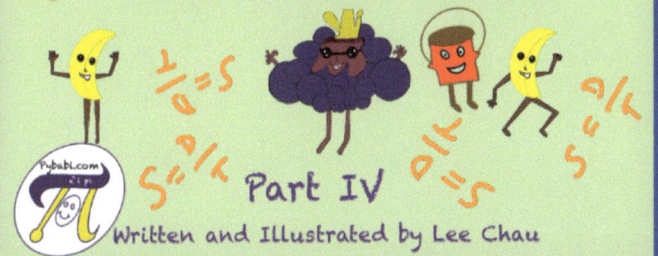

THE BIG RACE

Part IV

Written and Illustrated by Lee Chau

This is my book.

Date: _____

Name:

_____

I love you Space

Please sign, date and leave words of encouragement

I love you Space

Please sign, date and leave words of encouragement

I love you Space

Please sign, date and leave words of encouragement

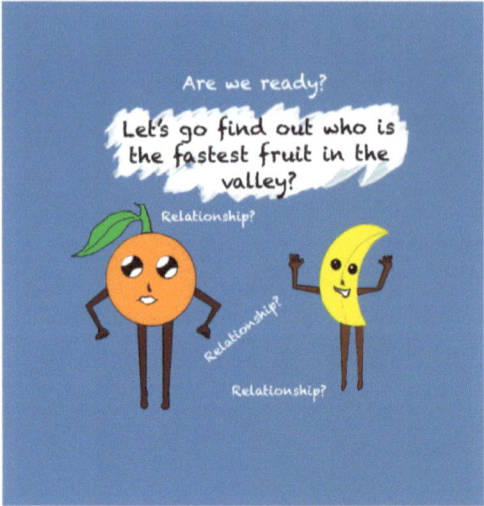

First we need to introduce you to 2 of our friends!

Secky

Disty

1

Secky and Disty are changing properties.

Let's watch and see how they come to an agreement.

I need rocks

I have buckets

2

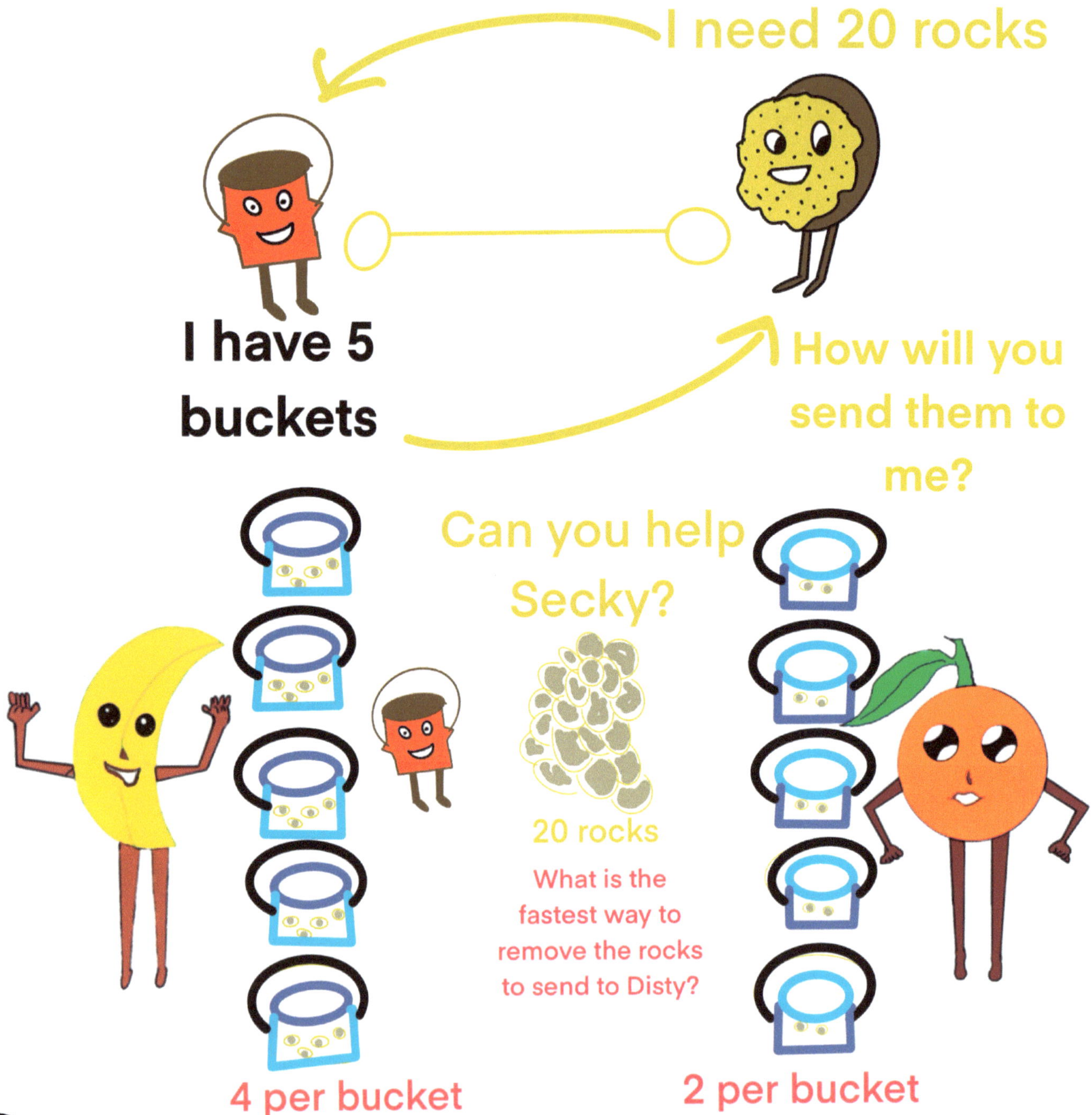
I need 20 rocks
I have 5 buckets
How will you send them to me?
Can you help Secky?
20 rocks
What is the fastest way to remove the rocks to send to Disty?
4 per bucket
2 per bucket

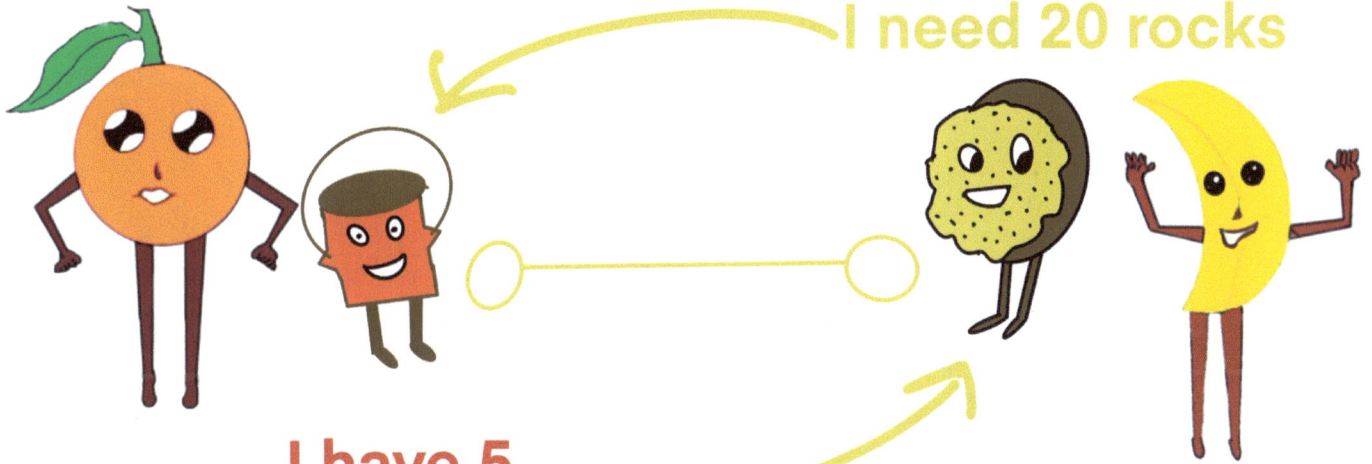

I need 20 rocks

I have 5 buckets

How will you send them to me.

I will send 4 rocks in each bucket

AGREED!

4

Can you see that it does not matter how big the pile of rocks I have to remove is or the amount Disty needs?

The agreement of 4 rocks per bucket will always be faster than 2 rocks per bucket with a limit of 5 buckets. If the amount of buckets are the same we can see which agreement was the fastest by seeing who transferred the most rocks after the fifth or last delivery. Whoever deposited the most rocks had removed the pile the fastest.

But to see "how much faster" we can look at the two agreements.

We only need to look at one bucket from each agreement because the total number of buckets are the same and then count the rocks.

1 bucket = 4 rocks

The second agreement is 1 bucket = 2 rocks. So 4 is twice as large as 2 so 4rocks/1bucket is twice the size of 2 rocks per bucket so the 4 to 1 agreement is 2 times faster than the 2 to 1 agreement.

MATCH 1

He shared 20 rocks into the 5 buckets. Same as: he divided 20 into 5 Same as: 20/5

What is the total measurement for Disty?

To send 4 rocks in each bucket Same as: 4 rocks per bucket Same as: 4 rocks/1 bucket Same as: 4/1

What is the agreement?

5

20

How did Secky determine what the agreement would be?

What is the total measurement for Secky?

6

MATCH 2

To look at the rocks shared into the buckets of each agreement to see which one is faster.

What does looking at rock to bucket numbers of each agreement tell you that the total delivered does not tell you?

1 rock per bucket

Yes

What does it mean to compare two agreement?

If two agreements have the same amount of buckets and one agreement moves 1 rock per bucket and the other moves .5 (a half of rock) per bucket which agreement is faster?.

one agreement is over the other agreement.

How much faster

fastest?

delivers the most amount of rocks the

If both agreements have the same amount of buckets, is the agreement that

7

MATCH 3

.83333333+.833
33333 = ?

Yes

Yes

2 sets of 2 fit
inside of 4.
2+2=4

1.666667

1.666666

How many sets
of 2 fit inside of
4?

Is 1 greater
than .83?

To remove 1.6666667
rocks per bucket is twice
faster than to remove
.83333333 rocks per
bucket?

8

ANSWER 1

What is the total measurement for Secky?

How did Secky determine what the agreement would be?

What is the agreement?

What is the total measurement for Disty?

ANSWER 2

If both agreements have the same amount of buckets, is the agreement that delivers the most amount of rocks the fastest?

What does looking at rock to bucket numbers of each agreement tell you that the total delivered does not tell you?

If two agreements have the same amount of buckets and one agreement moves 1 rock per bucket and the other moves .5 (a half of rock) per bucket which agreement is faster ?

What does it mean to compare two agreements?

10

ANSWER 3

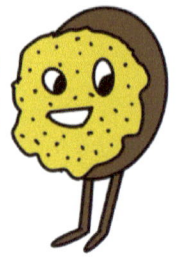

.83333333+.83333333 = ?

Is 1 greater than .83333333?

How many sets of 2 fit inside of 4?

Is to remove 1.66666667 rocks per bucket twice faster than to remove .833333333 rocks per bucket?

11

 # AGREEMENT 1

ratio =

unit change =

rate of change =

Incremental change =

each time of change =

12

AGREEMENT 2

Some more names for the agreement?

ratio

Derivative

rate of change

dy

slope

The letter 'm'

13

AGREEMENT 3

$$\frac{dy}{dx}$$

$f'(x)$

slope

m

S

 speed

V_{av}

rate of change

derivative

14

What's missing?

$$\frac{dy}{d}$$

$f'(\ )$

s ope

⊆

m

V v

pee

rate   change

d rivative

15

ANSWER 5

Tell us 8 ways to describe an agreement from AGREEMENT 3?

16

Tell us 8 ways to describe a

ratio from AGREEMENT 3?

1 changing property is located in one thing

The other changing property is in another thing

Looking at the ratio between 2 things.

18

Two changing properties are in one thing

Two changing properties are in the other thing

THING 1

THING 2

Looking at the ratio (derivative) in each thing and noting their differences.

19

Every moving thing has its own Secky and Disty. An agreement of changing properties that shows its ability to move through time and space.

THING 1

THING 2

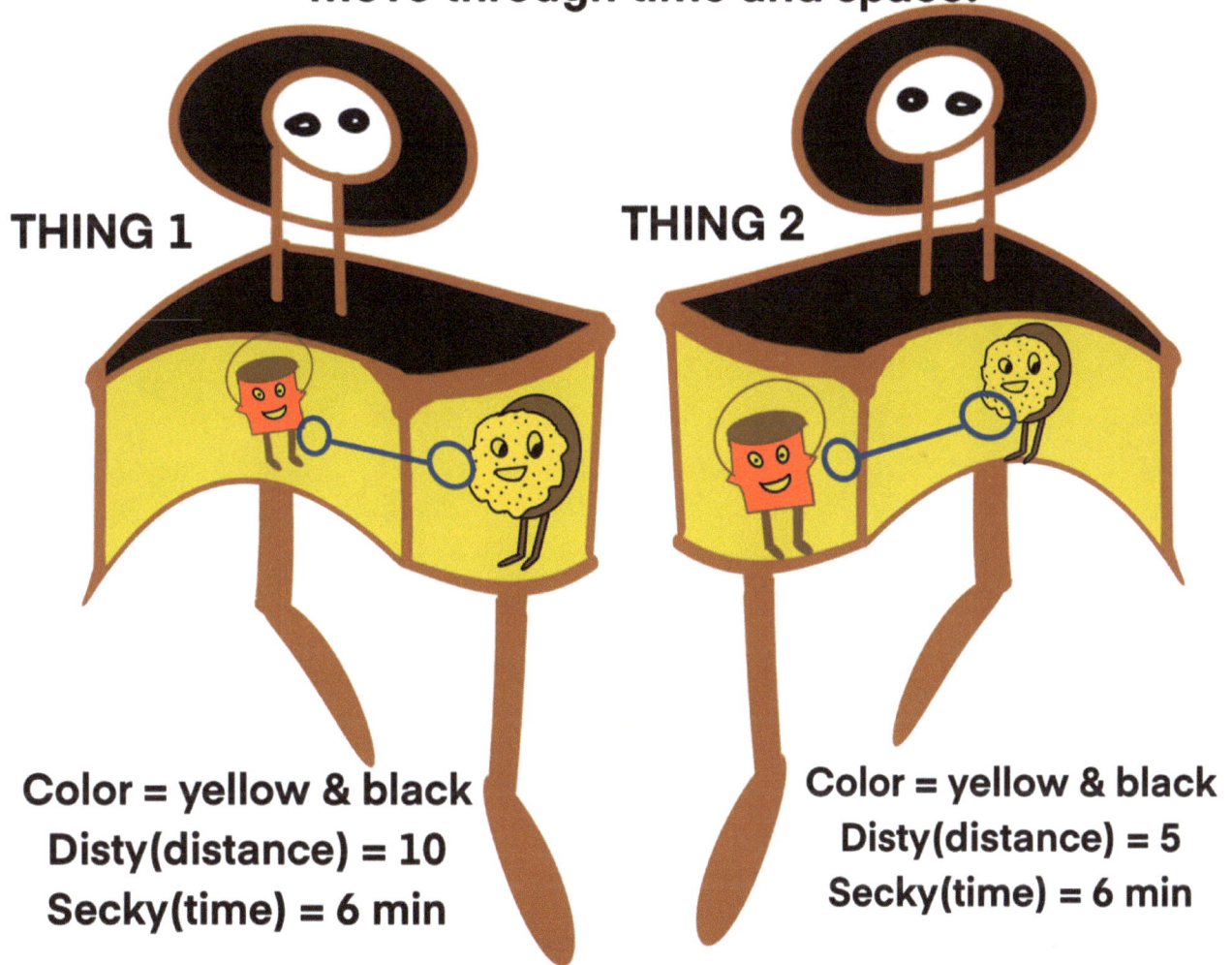

Color = yellow & black
Disty(distance) = 10
Secky(time) = 6 min

Color = yellow & black
Disty(distance) = 5
Secky(time) = 6 min

Which thing moves the fastest?

20

Just by looking we can already tell that 'thing1' is the fastest because it's total measure is greater than that of 'thing2'. The buckets (time) is the same for both runners. As we stated before the total measurement is not the most important thing. But rather the 'agreement (rocks per bucket) that is distance per unit of time.

THING 1

THING 2

Color = yellow & black
Disty(distance) = 10
Secky(time) = 6 min

Color = yellow & black
Disty(distance) = 5
Secky(time) = 6 min

THING 1 Ratio = 10/6 then 1.666666667 rocks per bucket
THING 2 Ratio = 5/6 then .8333333 per bucket

We can say that THING 1 is the fastest! It moves 1.66666667 per unit of time. While THING 2 moves only .833333333 per unit of time. This agreement (ratio, rate, derivative, Secky and Disty) that is in all things that move, is for short called SPEED! To remind us that we need to divide in order to see the agreement we use these letters.

$$S = \frac{D}{T}$$

$$SPEED = \frac{D}{T}$$

THING 1

Color = yellow & black
Disty(distance) = 10
Secky(time) = 6 min

22

FASTEST!

THING 1

Color = yellow & black
Disty(distance) = 10
Secky(time) = 6 min

SPEED

$$S = \frac{Distance\ Disty}{Time\ Secky}$$

THING 1

Color = yellow & black
Disty(distance) = 10
Secky(time) = 6 min

THING 1

Color = yellow & black
Disty(distance) = 10
Secky(time) = 6 min

THING 1

Color = yellow & black
Disty(distance) = 10
Secky(time) = 6 min

SPEED

$$S = \frac{Distance\ Disty}{Time\ Secky}$$

SPEED

$$S = \frac{Distance\ Disty}{Time\ Secky}$$

23

1 changing property is located in one thing

The other changing property is in another thing

Any changing property ACP

Any changing property ACP

Looking at the ratio between 2 things.

24

2 changing properties are located in one thing

2 changing properties are located in the other thing

We can look at both agreements (ratios) and see which ratio moves more rocks the fastest.

25

This arrangement of AGREEMENT can apply to more than just SPEED. It applies to any 2 changing properties. So, speed is just one type of derivative that applies to moving things. When the changing properties apply to something else we call the derivative by some other name.

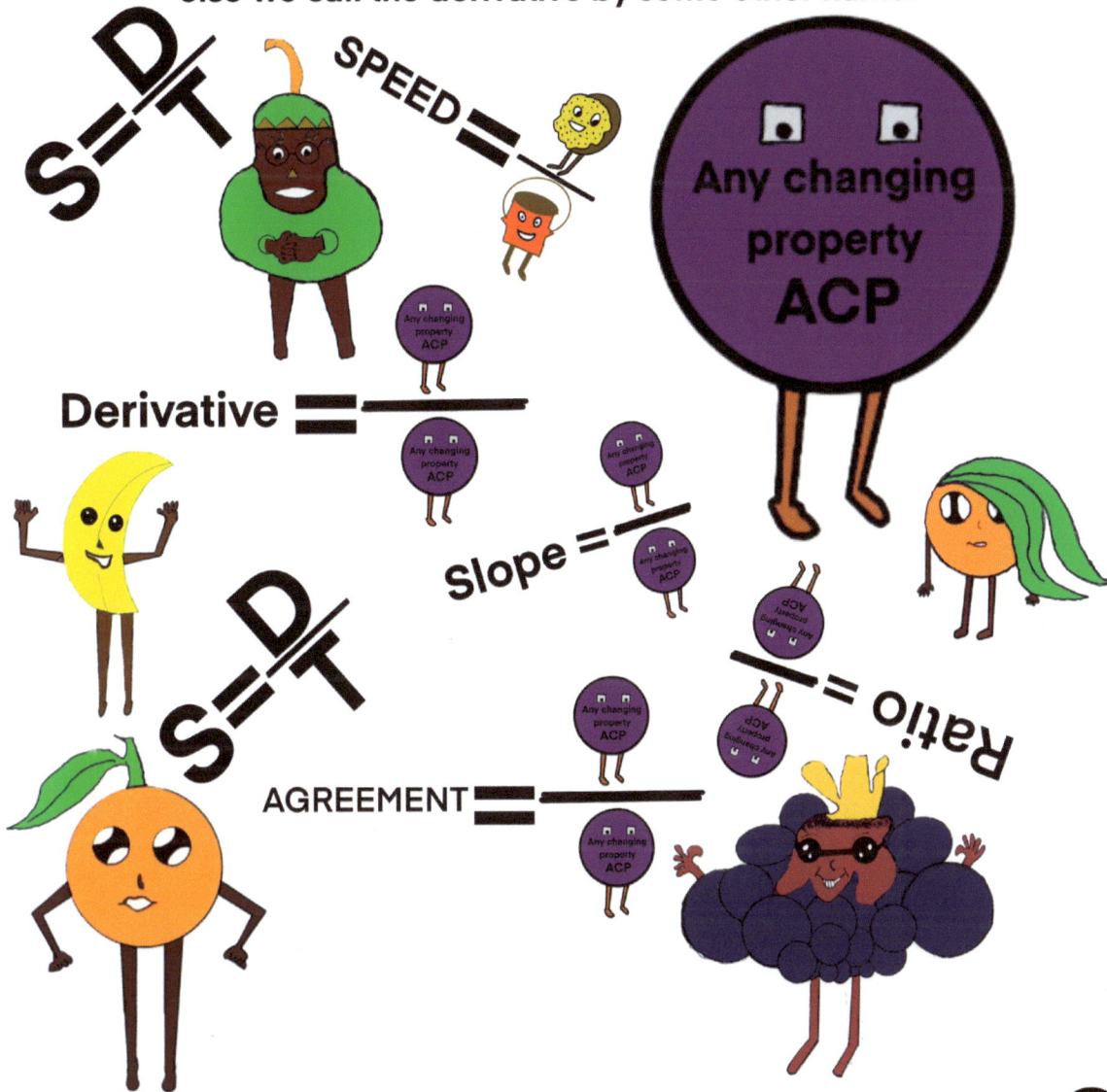

$$S = \frac{D}{T}$$

$$SPEED = $$

Any changing property ACP

$$Derivative = \frac{ACP}{ACP}$$

$$Slope = \frac{ACP}{ACP}$$

$$S = \frac{D}{T}$$

$$AGREEMENT = \frac{ACP}{ACP}$$

$$Ratio = \frac{ACP}{ACP}$$

26

What's inside a changing property that's located on the top?

This is me:

h = more buckets

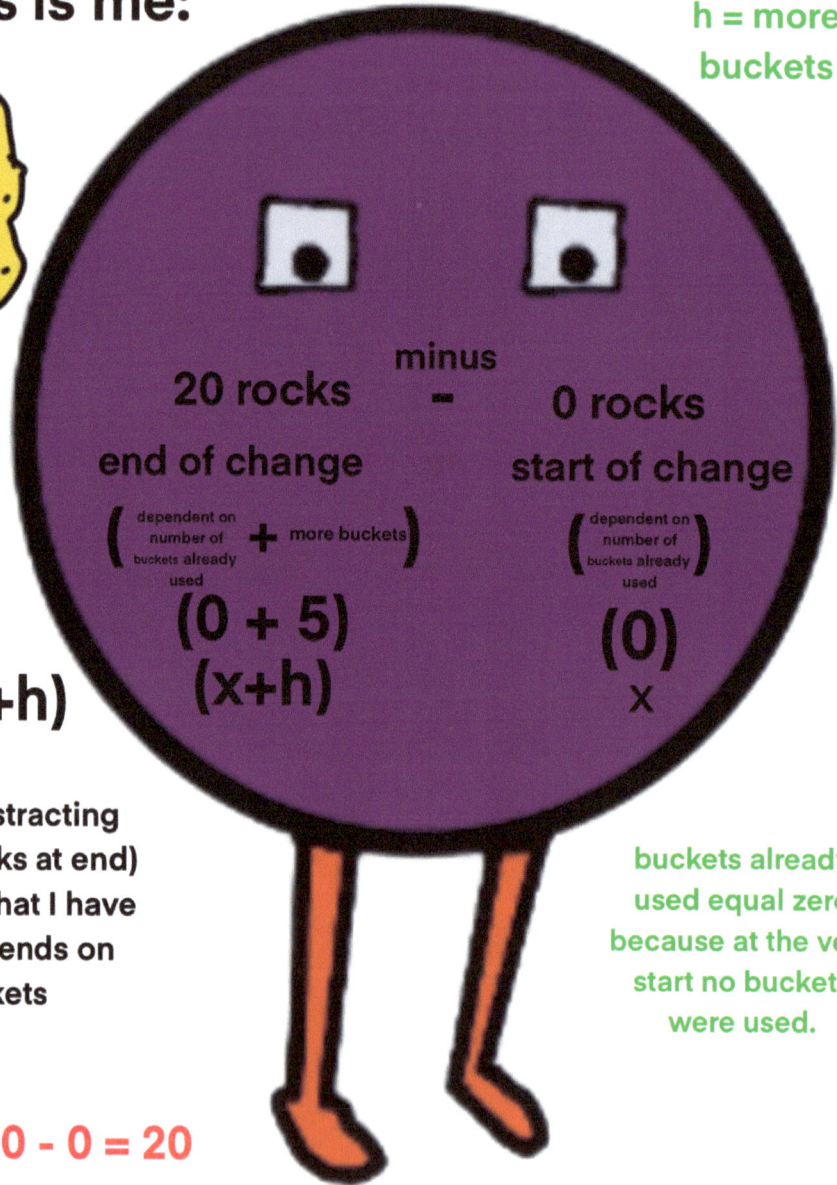

20 rocks
end of change

minus −

0 rocks
start of change

(dependent on number of buckets already used) **+** more buckets

(0 + 5)
(x+h)

(dependent on number of buckets already used)

(0)
x

buckets already used = zero. **= X**

buckets already used = zero. plus more buckets **= (x+h)**

I get my number by substracting (rocks at start) from (rocks at end) but the amount of rocks that I have at the start and end depends on Secky and the buckets

buckets already used equal zero because at the very start no buckets were used.

ANSWER:
Substraction 20 - 0 = 20

27

What's inside a changing property that's located on the bottom?

This is me:

h = more buckets

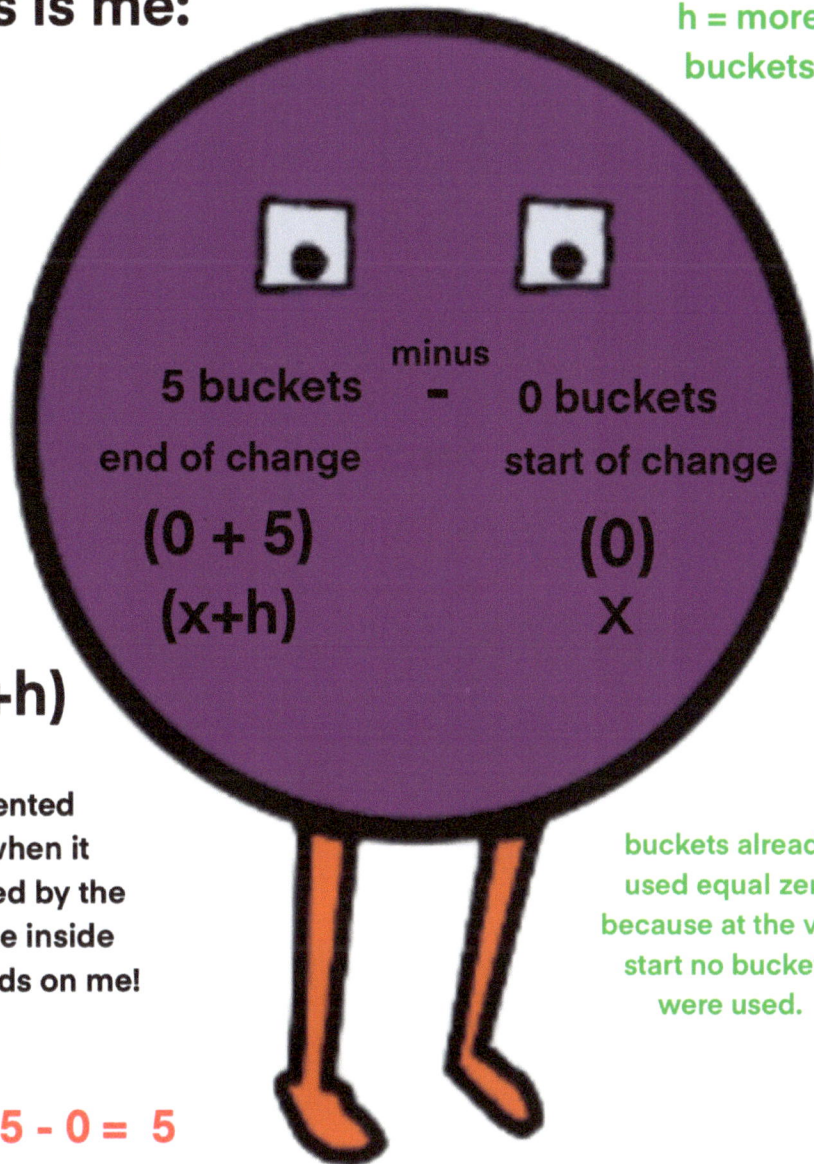

buckets already used = zero. **= X**

buckets already used = zero. plus more buckets **= (x+h)**

5 buckets
end of change

minus
-

0 buckets
start of change

(0 + 5)
(x+h)

(0)
X

buckets already used equal zero because at the very start no buckets were used.

My number is represented by the lettter 'X' and when it increases it is represented by the letter 'h'. You will see me inside Disty because she depends on me!

ANSWER:
   Substraction 5 - 0 = 5

28

We hope that these lessons have helped you to solve the mystery of 'Who is The Fastest Fruit in The Valley?'

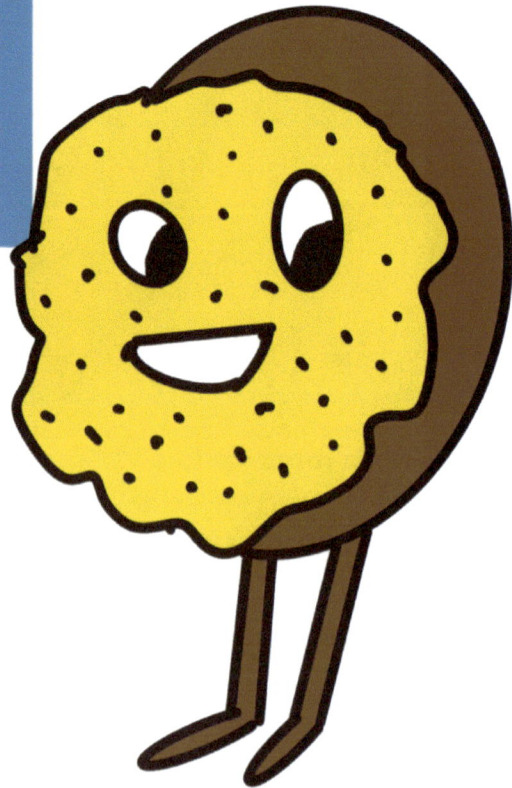

29

I am the fastest fruit in the valley!

My speed is like that of 'THING 1 and Banana Balii speed is like that of 'THING 2'

I ran 10 long 6 minutes

Balii Banana ran 5 long 6 minutes

My average speed was 1.6666667 per minute

Balii's average speed was .833333 per minute

I ran twice as fast as Balii.

30

Teachers Space

Please sign, date and leave words of encouragement

Friends Space

Please sign, date and leave words of encouragement

Mess up Space

Do whatever you want here.

THE BIG RACE

Part I

Written and Illustrated by Lee Chau

THE BIG RACE

Part II

Written and Illustrated by Lee Chau

THE BIG RACE

Part III

Written and Illustrated by Lee Chau

THE BIG RACE

Part IV

Written and Illustrated by Lee Chau

Let's go to part IV

Can you share your story?

Tiktok @ogi.orange

Twitter @ogi_orange

Instagram @ogi.orange

Email: ogiorange@pybabi.com

www.ingramcontent.com/pod-product-compliance
Lightning Source LLC
Chambersburg PA
CBHW041547040426
42447CB00002B/75